Diana Santos Sales

Therapeutic use of Pequi

Diana Santos Sales

Therapeutic use of Pequi

Botany, uses, chemical constituents and pharmacological aspects

ScienciaScripts

Cover image: www.ingimage.com

This book is a translation from the original published under ISBN 978-613-9-68274-4.

Publisher:
Sciencia Scripts
is a trademark of
Dodo Books Indian Ocean Ltd. and OmniScriptum S.R.L publishing group

120 High Road, East Finchley, London, N2 9ED, United Kingdom
Str. Armeneasca 28/1, office 1, Chisinau MD-2012, Republic of Moldova, Europe
Printed at: see last page
ISBN: 978-620-8-20569-0

SUMMARY

DEDICATORY 2
ACKNOWLEDGMENTS 3
Author's Profile 4
1 INTRODUCTION 5
2 OBJECTIVES 7
3 THEORETICAL FRAMEWORK 8
4 METHODOLOGY 18
5 Results and discussion 20
6 FINAL CONSIDERATIONS 32
REFERENCES 34

DEDICATORY

I dedicate this and all my other achievements to my beloved husband, who has given me strength and courage in a special and loving way, supporting me and being with me in times of difficulty.

"Persistence is the path to success."

Charles Chaplin

ACKNOWLEDGMENTS

First of all, I thank God for the gift of life and for enabling me at all times.

I would like to thank my husband Filipe, the main person responsible for me getting this far, for his patience and support, not letting me give up and always encouraging me to carry on.

I would like to thank my friends and colleagues who contributed in one way or another to this special moment.

I would like to thank my teachers, all those with whom I have spent the last five years and who have been part of this arduous journey at university.

I would like to thank my advisor for his trust, attention, patience and for helping me at every stage of this research.

AUTHOR'S PROFILE

Pharmacist. Postgraduate student in Clinical Pharmacy and Pharmaceutical Prescribing at

Institute of Science, Technology and Quality (ICTQ).

E-mail: dssa.diana@gmail.com

1 INTRODUCTION

The pequi is a typical species of the Brazilian cerrado, belonging to the genus *Caryocar brasiliense* of the *Caryocaraceae* family. The tree can reach a height of up to 12 meters. The pequi tree is one of the plants most commonly used to feed humans and is increasingly being featured in culinary dishes and is present on various gastronomic menus (CARVALHO; PEREIRA; ARAÙJO, 2015).

Numerous researchers have highlighted the pharmacological effects of parts of the pequi tree such as leaves, bark and the fruit itself (SILVA; DIAS; FIGUEIRINHA, 20il). It is a fruit that has numerous therapeutic properties and can also be used in the cosmetics industry, using the extraction of pequi oil (CARVALHO; PEREIRA; ARAÙJO, 2015). It is also rich in vitamins and essential oils that act on the bone, muscle, endocrine and immune systems (CAMPOS; GOMIDES; RIBEIRO; ARAÙJO, 2012).

In folk medicine, the leaves and fruit are used for respiratory diseases and the seeds are used as aphrodisiacs. The oil from the pulp has an invigorating medicinal effect and is used against bronchitis, flu, colds and has antifungal activity (CARVALHO; PEREIRA; ARAÙJO, 2015).

The pequi (*Caryocar brasiliense)* is a fruit that has different metabolic classes, besides being rich in lipids and fiber, it is also rich in vitamins, beta-carotenes, proteins and calcium. It is a fruit with lipids that facilitate the catalysis of other lipids and most of its lipids are soluble (SILVA; DIAS; FIGUEIRINHA, 2011).

We were interested in this area of study, researching through a bibliographical review the possibilities of using *Caryocar brasiliense* as a medicine because it has numerous therapeutic effects that make it the target of great research and experimentation due to the richness of its nutrients and possible pharmacological actions. An analysis was made of the knowledge regarding its use, which made it of immense importance and of great interest to the population in general (CAMPOS et al., 2012).

It is a fruit that plays an important role in generating jobs and family income for the agro-extractivist population. At harvest time, the sale of the fruit drives a large part of the economy in the cerrado region, earning profits (SANTOS et al., 2010).

Studies of the chemical characterization of this fruit show that both the pulp and the kernel of the pequi are rich in lipids, unsaturated fatty acids and saturated fatty acids. These compounds show that this food, consumed in moderation, can have wide-ranging benefits for human health (BORGES, 2011).

2 OBJECTIVES

2.1 General objective

To describe the possible uses of *Caryocar brasiliense* from a pharmacological perspective.

2.2 Specific objectives

Obtain information on the pharmacological use of *Caryocar brasiliense* through bibliographical research;

Characterize the history of use of *Caryocar brasiliense* in folk medicine;

Investigate the parts of the plant used with pharmacological properties.

3 THEORETICAL FRAMEWORK

3.1 Botanical description of *Caryocar brasiliense*

The pequizeiro (Figure 1) is a tree that grows to between 10 and 12 meters in height. It is one of the plants used to feed the people of the cerrado and is increasingly being featured on restaurant menus. It is an ornamental plant due to the shape of its crown and the external arrangement of its white flowers. It flowers from June to October and bears fruit from August to January and can be found until February (Figure 2) (BORGES, 2011; SANTOS et al., 2013).

The pequi (*Caryocar brasiliense*), also known according to its region of occurrence as piqui, piqui-do-cerrado, almond of thorn, belongs to the family *Caryocaraceae* of the genus *Caryocar brasiliense*, has about 20 species, 12 of which are found in Brazil. Pequi or piqui is a Tupi name where "py" means bark and "qui" means thorns, it is an arboreal plant with a wide distribution, found in several Brazilian states, Bahia, Cearà, Goiàs, Mato Grosso, Parà, Distrito Federal, São Paulo, Maranhao, Paranà and Minas Gerais (SILVA et al., 2010; CARVALHO et al., 2015.

The pequi tree has a stem with very resistant wood of excellent quality, globose in shape, with a light green color in which there are one to four "stones" that have thorns in the innermost layer. Its leaves are made up of trifoliate, opposite parts that can measure up to 20 centimeters in length and are covered in dense, bright green hair, with no hairs or glands (BORGES, 2011).

It is a fruit made up of a pericarp, mesocarp and endocarp made up of layers. The pericarp corresponds to the skin, which is greenish or grayish in color and is the smallest part of the fruit. The mesocarp, the part where the pulp is found, is made up of a whitish or light yellow layer, aromatic, fleshy, rich in tannin and is considered the edible part of the fruit; and the endocarp, the portion surrounding the stone and adhered to the mesocarp, is covered with thorns in one layer.

The innermost part is rigid and hard on the outside, 2 to 5 millimeters long and reddish in color, protecting the kernel or seed (CARVALHO et al., 2015).

Figure 1: Adult *Caryocar brasiliense* plant

Source: Carlos, 2012.

Figure 2: *Caryocar brasiliense* flower

Source: Borges, 2012

There are between 500 and 2,000 fruits on each plant, ranging from 6 to 14 centimeters in length and 6 to 10 centimeters in diameter. Their weight per unit varies from 100 to 300 grams, while the number of seeds can vary from one to four per fruit. They are white in color and have an oleaginous texture, in the shape of a small kidney (Figure 3) (BORGES, 2011; CARVALHO et al., 2015).

Figure 3: *Caryocar brasiliense* fruit

Source: Carlos, 2012.

3.2 Uses of *Caryocar brasiliense*

The pequi tree is considered a species of economic interest due to the use of its fruit in cooking, as a source of vitamins and in the extraction of oils for the manufacture of cosmetics. The fruit and leaves of *Caryocar brasiliense* are also used for therapeutic purposes (SANTOS et al., 2010; BORGES, 2011).

The bark of the pequi tree is used as a dye, providing a yellow-brown color. The kernel together with the pulp (mesocarp) is used in food, where it is cooked with rice, beans, chicken and other culinary dishes and can also be used to prepare liqueur and to extract butter. The oil extracted from the pequi pulp has a toning effect, as well as acting against colds and flu and controlling tumors. This oil is usually mixed with bee honey or capybara lard and used as an expectorant. It can also be used for burns and edema. Because it has various therapeutic properties such as vitamin A and C, thiamine, proteins and minerals, pequi oil is used in folk medicine to prevent eye problems which are related to vitamin A deficiency (BORGES, 2011; CAVALCANTI, 2012; SANTOS et al., 2010).

The fruits are also considered aphrodisiacs for men and fortifying for pregnant women (SANTOS et al., 2010).

It is a fruit that has a high content of carotene, a substance that has yellowish, orange and red pigments. They play an important role in cell differentiation and also act on the immune system, preventing infections (SANTOS et al., 2010).

The pequi pulp has a considerable percentage of pectin, which is important in food processing, as it provides firmness, retains flavor and aroma, making it very important for the food industry (CARVALHO et al., 2015).

The pequi kernel contains unsaturated fatty acids, such as oleic and palmitic acids, among other acids such as omega 3 and omega 6 found in high quantities. These fatty acids are the main source of energy reserves for human beings and have the function of preventing some chronic diseases, since they form part of cell membranes and are also precursors of substances such as prostaglandins, thromboxanes and leukotrienes (CAMPOS et al., 2012).

The part of the pequi that is used in both industry and cooking is the mesocarp, found inside the fruit, which is the fleshy part and is considered the edible part. It is composed internally of a sclerotic tissue deposited on the endocarp, where there are more thorns (SANTOS et al., 2010).

3.3 Main chemical constituents of Caryocar *brasiliense*

According to studies carried out by Dias et al. (2011), secondary metabolites were found in the phytochemical prospecting of *Caryocar brasiliense* leaves, in which these metabolites consist of steroids, triterpenoids, cardiotonic heterosides, flavonoids, tannins, alkaloids, coumarins and essential oils.

A phytochemical study of the hydroethanolic extract of pequi leaves found the presence of flavonoids, saponins, tannins and cardiotonic glycosides and the absence of alkaloids and anthraquinonic glycosides in the extract (CARVALHO et al., 2015).

According to these studies, condensed tannins, water-soluble tannins, flavonoids and terpenes have also been detected in the ethanolic extract of pequi leaves and bark. These are metabolite components found in various parts of the fruit with therapeutic properties for various treatments. Cardiotonic heterosides have properties in the treatment of Congestive Heart Failure (CHF) and also for the treatment of Cardiogenic Shock in cases of Pulmonary Edema. Care should be taken with regard to the toxicity of these components, since in high concentrations they can cause adverse reactions (CARVALHO et al., 2015; (CAMPOS et al., 2012).

The leaves of *Caryocar brasiliense* contain a high amount of phenolic compounds, such as flavonoids, which are of great pharmacological interest. Among these pharmacological interests are therapeutic properties such as antispasmodic, antiallergic, antiviral, anti-inflammatory and antiulcerogenic properties (CAMPOS et al., 2012).

Phenolic compounds are included in the category of free radical blockers, in which they are used as antioxidants and can reduce oxidative damage in humans (CARVALHO et al., 2015).

The oil from the pulp has also shown invigorating properties and can act against bronchitis, colds, flu and to control tumors. The oil content and antioxidant compounds in pequi seeds, pulp and flowers are of great interest to the pharmaceutical industry, as well as in the manufacture of cosmetics, since the oil extracted from the seeds has a mild smell and a pleasant texture, and is used in the manufacture of creams and soaps for aesthetic use, with pequi as the main component (SANTOS et al., 2010; CARVALHO et al., 2015).

Another metabolic group found in the leaves of *Caryocar brasiliense* are tannins. Studies carried out on pequi leaves and stems have revealed the presence of condensed and hydrolysable tannins. Belonging to the group of secondary metabolites, tannins are of great economic interest, since the intake of diets rich in fruits containing the compound is associated with anticarcinogenic activity. In addition, tannins can act as an anti-inflammatory and healing agent (CARVALHO et al., 2015).

Tannins are present in pequi oil, which gives it healing activity. They are compounds that have the ability to precipitate proteins and sequester metal ions, especially iron, which is essential for the development of microorganisms, providing an antimicrobial and antifungal effect (BATISTA et al., 2010).

In the leaves of *Caryocar brasiliense*, according to studies carried out by Dias (2010), coumarins were found, which are attributed with a wide range of biological activities. Coumarins have therapeutic actions such as antimicrobial, antioxidant, antiviral and antitumor.

Among these studies carried out by various authors on the active principles of *Caryocar brasiliense,* it was also possible to identify the presence of alkaloids in pequi leaves, since alkaloids are examples of secondary metabolites that have the function of originating various drugs, making their structural diversity and pharmacological activities, constituent metabolite groups of great importance among natural substances with therapeutic interest. Alkaloids are substances from secondary metabolism that stand out for their intense action on the central nervous system (CNS), many of which are used as poisons or hallucinogens (CARVALHO et al., 2015).

Other chemical compounds also found in the leaves of *Caryocar brasiliense* were steroids, according to studies carried out by Dias and Morais (2011). Plants that contain steroids are of interest to the pharmaceutical industry because they are substances that represent starting molecules for the semi-synthetic production of steroidal drugs, such as contraceptives, steroidal anti-inflammatory drugs and anabolic steroids (CARVALHO et al., 2015).

3.4 Pharmacological aspects of Caryocar *brasiliense*

Despite few scientific studies on its biological activities and ethnobotanical uses, pequi has gained the attention of researchers due to its antibacterial, antifungal, parasiticidal and antioxidant therapeutic activities (BORGES et al., 2011; CARVALHO et al., 2015).

The fruit of the pequi tree (*Caryocar brasiliense*) contains around 33% lipids, most of which are oleic acid, an unsaturated fatty acid that is essential for the animal organism. Among the studies carried out, the anti-tumor effect stands out. The isolated effect of the ethanolic extract of pequi leaves evaluated on mice with sarcoma 180 in solid form showed an inhibitory effect of between 15 and 67% on the development of the tumor, which may have antitumor properties (CARVALHO et al., 2015).

According to the researcher Santiago (1998), he isolated and characterized proteins from the fruit extract using the epicarp, mesocarp and endocarp, of which only the endocarp extract promoted a reduction in the viability of tachyzoite forms of toxoplasma gondii (SANTOS et al., 2010).

According to studies carried out by Passos (2001), the antifungal activity of pequi on isolates of *Cryptococcus neoformans Paracoccidioides brasiliensis* was evaluated. The toxicity of antifungal agents and the emergence of microbiological resistance have led to the need to develop new antifungals that have greater advantages over existing drugs (SANTOS et al., 2010).

It was found that all parts of the fruit had antifungal activity, with the wax removed from the leaves having the highest activity, inhibiting the growth of 91.3% of *Cryptococcus neoformans* isolates (SANTOS et al., 2010; CARVALHO et al., 2015).

In addition to these findings, recent studies show that *Caryocar brasiliense* has methanolic and ethanolic extracts of leaves, flower buds and fruits, which include the outer and inner mesocarp and almonds, with a toxic effect against the germination of spores of some fungi, including *Botrytis cinerea, Fusarium oxyporum* (CARVALHO et al., 2015).

In the ethanolic extract of pequi leaves and bark, a concentration of 100 ppm was found to have molluscicidal activity against *Biomphalaria glabrata*, in which the death of 90% of the intermediate host of *Schistosoma mansoni*, the etiological agent that causes the disease schistosomiasis, was observed (BORGES, 2011; CARVALHO et al., 2015).

Studies carried out by Herzog et al. (2002) evaluated the effect of the crude ethanolic extract of the bark of *Caryocar brasiliense* at 400 ppm in mice infected intraperitoneally with trypomastigote forms of *Trypanosoma cruzi*, in which a significant interference in the parasitemia curve was observed, compared to the control group, with a reduced number of parasites in the blood.

According to studies by PAULA Júnior et al. (2006), the hydroethanolic extract of *Caryocar brasiliense* leaves interferes with the growth of different groups of pathogenic microorganisms in humans. The presence of antimicrobial action was observed against strains *of Enterococcus faecalis, Escherichia coli* and *Staphylococcus aureus*, which were evaluated using the agar diffusion method. The hydroethanolic extract at concentrations of 2.5 and 5.0 mg/mL inhibited the proliferation of promastigote forms of *Leishmania amazonensis*, and the effect was found to be superior to the effects of meglumine antimoniate (Glucantime), the drug of choice for the treatment of leishmaniasis (CARVALHO et al., 2015; BORGES, 2011).

The studies carried out by Miguel et al. (2011) evaluated the neuroprotective action of the ethanolic extract of the bark of *Caryocar brasiliense* in the brains of rats subjected to ischemia and reperfusion lesions and found that the ethanolic extract of the bark of *Caryocar brasiliense* at higher concentrations of 300 and 600 mg/Kg shows possible neuroprotective activity, due to a reduction in the number of ischemic neurons in the frontal cortex region (CARVALHO et al., 2015).

Among the studies carried out, Roesler et al. (2010) evaluated the cytotoxicity and photoxicity *in vitro* of the ethanolic extract of the bark of *Caryocar brasiliense* and observed that at concentrations of up to 3000 μg/mL[1] a precipitate was formed in the cell culture medium, thus causing a decrease in cell viability, while not showing phototoxic potential.

Some *in vitro* studies have evaluated the efficacy of the aqueous extract of *Caryocar brasiliense* pulp against mutagenicity. The results suggested that the antimutagenic potential was possibly due to the antioxidant properties of *Caryocar brasiliense* (BORGES, 2011).

In view of the various studies carried out, there is still a lack of scientific evidence in the literature about the therapeutic efficacy of *Caryocar brasiliense* in terms of its biological activities, the parts of the fruit and its toxicity index. Scientific support is needed to expand knowledge about the therapeutic effects, toxicity and adverse effects of *Caryocar brasiliense* (BORGES, 2011).

4 METHODOLOGY

For Fonseca (2002) methods means organization, and logos, systematic study, research, investigation, in other words, methodology is the study of organization, of the paths to be taken in order to carry out research or a study, or to do science. Etymologically, it means the study of the paths, the instruments used to carry out scientific research.

Therefore, the purpose of the methodology is to analyze and provide researchers with characteristics of the various methods available, thus evaluating their capacity as well as their potential during the investigation process for the research work.

4.1 Type of research

This work is a bibliographical review, which according to Marconi and Lakatos (2005), covers all the literature already published on the subject of study, consisting of books and scientific articles. Its purpose is to bring the researcher into direct contact with all the written material on a specific subject.

According to Gil (2012), the main advantage of bibliographical research lies in the fact that it allows the researcher to cover a much wider range of phenomena than they could research directly.

According to Tozoni (2006), although bibliographical research is a very particular type of research, we are not going to "listen" to interviewees or observe lived situations, but rather "listen", talk and debate with the authors through their writings.

As for the approach, this is qualitative research, which consists of presenting the information collected using an inductive approach. According to Gil (2010), qualitative research considers that there is a dynamic relationship between the real world and the subject, i.e. an inseparable link between the objective world and the subject's subjectivity that cannot be translated into numbers.

Qualitative research therefore refers to research that does not require statistical techniques or methods, and its data is analyzed and collected inductively. In terms of objectives, this is a descriptive study, the aim of which is to produce in-depth information based on an analysis of existing data, thus enabling new information to be gathered.

The bibliographical review was carried out by consulting scientific articles available on the Internet.

4.2 Data collection and processing

The bibliographic research was based on a survey of theoretical references related to the content. The searches were carried out using written and electronic databases, such as books, scientific articles, web pages, such as Bireme, Scielo, Google academic from the sources Lilacs and Pubmed with a year limit between 2010 and 2016 published in Portuguese and without restriction to the type of publication, using the following descriptors: pequi, medicinal use, folk medicine.

5 RESULTS AND DISCUSSION

In the first phase of this study, 39 scientific articles were found. After analyzing all this material, it was found that 11 manuscripts met the criteria established by the article's objectives, thus constituting the study's final sample. The analysis of the empirical material involved in this investigation made it possible to characterize the scientific productions included in the study, as shown in Flowchart 1 below.

The analysis included scientific articles published in the following databases: Bireme, Google Scholar, Scielo and Lilacs. In Google Scholar, 16 articles were obtained, of which two were selected according to the inclusion criteria and 14 were excluded. In Scielo, 16 articles were obtained in total, of which five were selected and 11 excluded; and in Lilacs seven articles were selected, but of these, four were selected and three were excluded. On other academic database sites such as Bireme, one article was found, which was not selected and excluded. And no articles were found on other websites. In the final phase, 11 articles were selected for full content analysis and included in the integrative review (Flowchart 1).

Flowchart **1**- Flowchart of literature identified and selected according to database

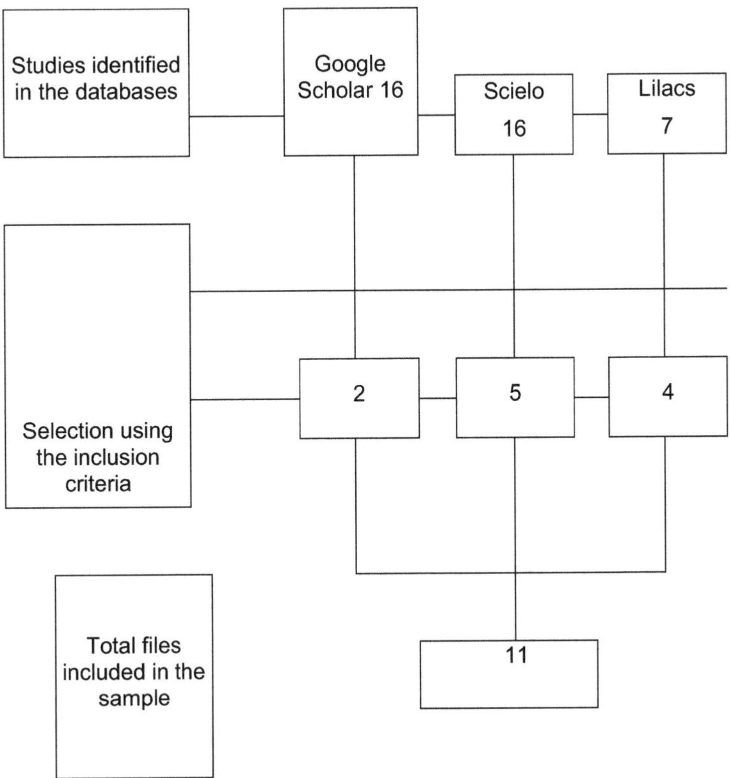

Table 1 shows the distribution of the 11 articles listed in the Google Scholar, Scielo and Lilacs databases, as well as their cataloging.

Table 1: Scientific publications on the possibilities of using *Caryocar brasiliense* from a pharmacological perspective, according to author, title, objectives and main results.

Author(s)	Year	Title	Objective	Results
BATISTA, J. S.; SILVA, A. E.; RODRIGUES , C.M.F.; COSTA, K.M.F.M.; OLIVEIRA, A.F.; PAIVA, E.S.; NUNES, F.V.A.; OLINDA, R.G.	2010	Evaluation of the healing activity of pequi oil (*Caryocar brasiliense*) on experimentally produced skin wounds in rats	To demonstrate the effects of topical treatment of pequi oil-based cream (*Caryocar brasiliense)* using 40 male Wistar rats aged 60 days.	The use of *Caryocar brasiliense* pulp oil had a positive influence on the healing of experimental skin wounds in rats, as it promoted a less intense inflammatory reaction and faster wound closure.
DINIZ, D. M.	2015	Anti-inflammatory activity of microemulsion containing pequi oil (C*aryocar brasiliense*)	To develop a macroemulsion for local topical administration, through the elaboration of a	The biological study revealed an increase in the activity of pequi oil when incorporated into

			pseudothermal phase diagram (DFPT), and to apply it in an inflammatory model.	the microemulsion and these results collaborated with the histological analysis carried out. This nanosystem is therefore presented as
				a promising carrier to improve anti-inflammatory activity .
BORGES, J. C. A.	2011	Botanical characteristics, nutritional aspects and therapeutic effects of pequi (*Caryocar brasiliense*)	Present what has been discovered and scientifically proven about the therapeutic activity of pequi and its derivatives.	The pulp and bark of pequi have significant amounts of phenolic compounds, which are potential

				antioxidants.
MOURA, L. R; MARTINS, A. C.; VAZ, L. A. R.; ORPINELLI, S. R. T.; SILVA, T. L.; FALEIRO, M. B. R.; SANTOS, S. C.; MOURA, V. M. B. D.	2012	Hydroalcoholic extract of pequi bark (*Caryocar brasiliense*) in rats submitted to Doxorubicin application	To evaluate the antioxidant effect of the hydroalcoholic extract of pequi bark (EHCP) in rats after administration of doxorubicin (DOX).	The use of the hydroalcoholic extract of pequi bark showed a possible antioxidant effect.
CAMPOS, D. G.; GOMIDES, J. N.; RIBEIRO, K. F. D.; ARAÙJO, S. C. M.	2012	Pequi: a proposal for teaching chemistry in secondary schools.	Bibliographical research into the chemical composition of pequi and its therapeutic properties, which were related to chemical content.	Studies of the chemical characterization of this fruit show that both the pulp and the kernel of the pequi are rich in lipids, unsaturated fatty

				acids and saturated fatty acids. There are countless therapeutic properties and benefits for human health.
BEZERRA, N.K.M.S.; BARROS, T. L.; COELHO, N.P.M.F.	2015	The action of pequi oil (*Caryocar brasiliense*) on the healing process of skin lesions in rats.	To analyze the effect of pequi oil on the healing process of skin lesions in rats.	The use of pequi oil had a positive influence on the repair process of skin lesions in rats, as evidenced by the faster closure of the wounds and the observation of inflammatory characteristics.
OAK, L. S.; PEREIRA, K. F.; ARAÙJO, E.	2015	Botanical characteristics, therapeutic effects and	Present the main active ingredients already identified	Studies have shown that pequi has several

G.		active ingredients present in pequi (*Caryocar brasiliense*)	in *Caryocar brasiliense* and also what has been discovered and scientifically proven about the therapeutic activity of this fruit.	chemical components, including phenolic compounds, which are known pharmacologically for their antioxidant potential.
CONCEIÇÃO ; G. M.; RUGGIER, A. C.; ARAÙJO, M. F. V. T.; CONCEIÇÃO , M. F. V; CONCEIÇÃO , M. A. M. M.	2011	Plants from the cerrado: commercialization, use and therapeutic indications provided by raizeiros and vendors, Teresina, Piaui	To get to know the species of medicinal plants from the Cerrado that are sold by raizeiros and vendors in the municipality of Teresina (PI).	Of all the species mentioned in this article that have pharmacological properties, *Caryocar brasiliense* stands out for its anti-tumor activities.
SANTOS, M. S; CAVALCANT I, D. S. P.	2010	The medicinal use of pequi (*Caryocar brasiliense)*	To provide greater knowledge about the medicinal use of *Caryocar*	There are several plants with therapeutic potential, among which *Caryocar*

			brasiliense	*brasiliense* stands out for its anti-inflammatory action. .
SILVA, A. A. A.; DIAS, J. A.; FIGUEIRINHA, M. O.;	2010	Benefits of the Brazilian pequi (*Caryocar brasiliensis*), a fruit native to the cerrado regions of the center-west.	Possibility of studying the pequi tree and publicizing the daily benefits of its consumption and pharmacological effects.	The results obtained in the experiments and biochemical and pharmacological studies of pequi point to its efficacy in the treatment of cancer, as well as for antifungal and bactericidal therapeutic purposes.
NTOS, F. S.; SANTOS, R. F.; DIAS, P. P.; JUNIOR, L.	2013	Pequi cultivation (*Caryocar brasiliense*)	Present a national overview of pequi cultivation and analyze its	The oil extracted from pequi has several applications in the

A. Z.; TOMASSONI , F.			therapeutic importance.	food and cosmetics industries, as well as therapeutic applications, such as the oil's antifungal activity and its effect on reducing inflammatory processes and blood pressure in runners.

Source: The author

The anti-inflammatory and healing action of pequi has been proven by several authors. Batista et al (2010) carried out a study on the healing evaluation of the plant, the subject of this work, using an animal model (rats) using the oil as a therapeutic agent and found that on the seventh day of treatment the wounds treated with the oil were covered with a thin crust, flush with the skin and without evidence of inflammation, however, these changes were not observed in the same period in the control group, whose wounds remained hyperemic, with swollen edges and purulent exudate.

The aforementioned researchers concluded that the analysis of the healing process from a clinical, macroscopic and histological point of view of the use of pequi fruit pulp oil had a positive influence on the healing of experimental skin wounds in rats, as it promoted a less intense inflammatory reaction and faster wound closure compared to the control group.

Recent studies by Bezerra, Barros and Coelho (2015) also had similar results to those of Batista et al (2010) and also reported that the animals treated with pequi oil had a greater stimulus for angiogenesis when compared to the control group. These researchers state that the use of pequi oil plays a beneficial role in tissue repair, as it promotes faster repair, as evidenced by faster wound closure and reduced inflammatory characteristics in the treated group compared to the control group.

Diniz (2015) showed in his research that the application of pequi oil to wounds can have its pharmacological effects optimized when it is prepared pharmacotechnically in the form of an emulsion. For the researcher, this pharmaceutical form proved to be a promising nanocarrier for the plant's oil, with a notable improvement in anti-inflammatory activities when compared to the oil applied purely.

The pharmacological actions of pequi seem to be as diverse as possible. Borges (2011) mentions that the consumption of pequi has been associated with a lower incidence of and mortality from various chronic non-communicable diseases. The protection that these foods offer against ischemic, cardiac and cerebrovascular diseases and cancer is associated with chemical constituents with antioxidant properties. The author has reported the use of the ethanolic extract of pequi leaves and bark on the death of 90% of a population of *Biomphalaria glabatra*, an intermediate host of *Schistosoma mansoni*, the etiological agent of schistosomiasis. The anti-parasitic actions don't stop there: the effect of the ethanolic crude extract of pequi bark on mice infected intraperitoneally with trypomastigote forms of the

Trypanosoma Cruzi strain on the eighth day showed significant interference with the parasitemia curve, compared to the control group, with a reduced number of parasites in the blood. Similar effects were detected in the promastigote forms of *Leishmania amazonensis*.

Carvalho, Pereira and Araùjo (2015) carried out a survey on the therapeutic use of pequi and reported that it has antifungal activity against *Cryptococcus neoformans*. They found that all parts of the fruit had antifungal activity, with the wax removed from the leaves having the highest activity, inhibiting the growth of 91.3% of the microorganism's isolates. They also described that the methanolic and ethanolic extracts of leaves, flower buds, fruit (outer mesocarp, inner mesocarp and kernel) also have a toxic effect on the germination of *Botrytis cinerea, Colletotrichum truncatum and Fusarium oxysporum* spores.

For the above authors, the extract of the leaves of the plant under study shows antimicrobial action against strains *of Enterococcus faecalis, Escherichia coli, Pseudomonas aeruginosa and Staphylococcus aureus*, evaluated by the agar diffusion method.

According to Conceiçâo et al (2010), the pharmacological activities of pequi are as varied as possible after carrying out a survey on marketing, use and therapeutic indications provided by raizeiros and vendors in Teresina, Piaui, and found that the use of its pulp has cured bronchitis, flu, colds and even controlled tumors. These researchers call for systematic studies to highlight the richness of the medicinal flora and, at the same time, identify the degree of threat caused by the pressure of collection on various species.

The toxic effects on parasites can be extended to animals and even humans. Moura et al (2012) evaluated the toxic action of the hydroalcoholic extract of pequi bark on rats and observed that under the conditions evaluated it increased the mortality rate in rats, which may be related to the toxic substances present in the bark of the fruit (SANTOS et al., 2013).

Another fact that caught the attention of this literary survey was the discourse by Silva, Dias and Figueirinha (2010) on the ethnobotanical knowledge of pequi for the treatment of cancer, as well as antifungal and bactericidal therapeutic purposes, but that this plant has been suffering from the action of man, since it has been used for other purely economic purposes. The authors report that the wood of the pequi tree has been used to make charcoal and it is known that it "has more value standing up than in a bag of charcoal" (GRISÓLIA, 2004). The cerrado has been destroyed by agriculture, but as well as the pequi there must be many other species that have not yet been properly studied.

The felling of the pequi tree could be avoided if man made use of other parts of the plant which have various medicinal and economic purposes, such as the oil extracted which has various applications in the food, pharmaceutical and cosmetics industries. Another application would be in the production of biodiesel, which highlights the importance of introducing this new source of energy.

6 FINAL CONSIDERATIONS

At the end of this study, it was revealed that the possibilities for using *Caryocar brasiliense* from a pharmacological perspective are as varied as possible, ranging from anti-inflammatory, antiparasitic, antifungal, antitumor and bactericidal actions.

It is a typically Brazilian fruit, used as a herbal medicine in popular medicine, which has become the target of researchers due to its pharmacologically discovered effects.

The most commonly used parts of the plant are the leaves and especially the fruit, which is rich in oil and can be made into a pharmaceutical emulsion.

The popular use of pequi is widespread throughout Brazil, especially in the cerrado, north and northeast regions of the country, whether in medicine or food.

Care should be taken with regard to the toxicological aspects of using this plant, as lethal actions have been reported in experiments, although it is not clear which chemical constituents confer such effects.

Therefore, it is believed that the objectives were met, as the importance of this fruit in the daily life of the cerrado population was verified, and it was noticeable that it had numerous effects on human health, and the results obtained in the experiments and pharmacological studies were satisfactory.

Caryocar brasiliense is of great importance to the agro-extractivist market and to the local economies of the cerrado region, since it is a fruit that is important for generating jobs and profits during harvesting seasons.

In folk medicine, its use is constant when the pulp is used to cure bronchitis, flu and colds. It can also be used as an expectorant when mixed with animal fat for possible treatment of burns and oedema.

It is hoped that this work will contribute to future research in order to provide even more in-depth knowledge on the subject, with more support on the toxicological effects, in order to bring more safety to the population.

REFERENCES

AGRA, M. F.; FREITAS, P. F.; BARBOSA, J. M. F. Synopsis of the plants known as medicinal and pois onous in North east of Brazil. **Braz. J. Pharmacog**. v. 17, n.1, p. 114-140, 2007.

ALMEIDA, A. C.; SOBRINHO, E. M., PINHO, L.; SOUZA, P. N. S.; MARTINS, E. R.; DUARTE, E. R.; SANTOS, H. O.; BRANDI, I. V.; CANGUSSU, A. S.; COSTA, J. P. R. Acute toxicity of hydroalcoholic extracts of rosemary, aroeira and barbatimâo leaves and pequi bark bran administered intraperitoneally. **Revista Ciência Rural**. Santa Maria, v. 39, 2009.

ALMEIDA, S. P.; SILVA, J. A. **Piqui e buriti**: importância alimentar para a populaçâo dos cerrados. Planaltina: EMBRAPA-CPAC, 1994. 38 p.

ALMEIDA, S. P. et al. **Cerrado: useful plant species.** Planaltina: EMBRAPA-CPAC, 1998. 464 p.

ASCARI, J.; TAKAHASHI, J. A.; BOAVENTURA, M. A. D. Phyto chemical and biological investigations of *Caryocar brasiliense* Camb. **Blacpma.** v. 9, n. 1, p. 20-28, 2010.

BATISTA, J. S. Avaliaçâo da atividade cicatrizante do óleo de pequi (Caryocar coriaceumWittm) em feridas cutâneas produzidas experimentalmente em ratos.**Arq. Inst**. Biol. v. 77, n. 3, p. 441-447, 2010.

BARRADAS, M. M. Morphology of the fruit and seed of *Caryocar brasiliense* (piqui) at various stages of development. **Revista de Biologia**, [S.l.], v. 9, p. 69-84, 1973.

BARREIRO, E. J.; FRAGA, C. A. M. **Quimica medicinal: as bases moleculares da açâção dos fàrmacos**. Porto Alegre: Artmed, 2001.243 p.

BEZERRA, J. C. B. et al. Molluscicidal lactivity against *Biomphalaria glabrata* of Brazilian Cerrado medicinal plants. **Fitoterapia,** v. 73, n. 5, p. 428-430, 2002.

BEZERRA, N.K.M.S.; BARROS, T. L.; COELHO, N.P.M.F.; The action of pequi oil (Caryocar brasiliense) on the healing process of skin lesions in rats. **Rev. Bras**. Pl. Med., Campinas, v.17, n.4, supl. II, p.875-880, 2015.

BRAGA, R. **Plantas do Nordeste, especialmente do Cearâ**. 2 ed. Fortaleza: Impressa Oficial, p.540,1960.

BRANDÂO, M.; LACA-BUENDiA, J. P.; MACEDO, J. F. **Arvores nativas e exóticas do Estado de Minas Gerais**. Belo Horizonte: EPAMIG, 2002. 528p.

BORGES, J. C. A. **Botanical characteristics, nutritional aspects and therapeutic effects of pequi (Caryocar brasiliense).** Seminars

Applied Program (Postgraduate). 2011.31p. Federal University of Goiás, Goiânia, 2011.

CAMPOS, D. G.; GOMIDES, J. N.; RIBEIRO, D. F.; ARAÙJO, S. C. M. **Pequi: a proposal for teaching chemistry in high school**. 2012. 12 p.

Chemistry Teaching Division of the Brazilian Chemical Society, Salvador, 2012.

CARVALHO, L. S.; PEREIRA, K. F.; ARAÙJO, E. G. Botanical characteristics, therapeutic effects and active principles present in pequi (Caryocar brasiliense). **Arq. Ciênc.** Saùde UNIPAR, Umuarama, v. 19, n. 2, p. 147-157, 2015.

CAVALCANTI, M. C. B. T.; RAMOS, M. A.; ARAÙJO, E. L.; ALBUQUERQUE, U. P.**Implications from the Use of Non-timber Forest Products on the Consumption of Wood as a Fuel Source in Human-Dominated Semiarid Landscapes.** Environmental Management, 2015.

CHÉVEZ - POZO, O. V. **The pequi (Caryocar brasiliense): an alternative for the sustainable development of the cerrado in northern Minas Gerais**. 1997. 97 p. Dissertation (Master's Degree in Rural Administration) Federal University of Lavras, Lavras, 1997.

CONCEIÇÂO; G. M.; RUGGIER, A. C.; ARAÙJO, M. F. V. T.; CONCEIÇÂO, M. F. V; CONCEIÇÂO, M. A. M. M. Plantas do cerrado: comercializaçao, uso e indicação terapêutica fornecida pelos raizeiros e vendedores, Teresina, Piaui. **Scientia Plena**. 2011.

CORDELL, G. A.; QUINN-BEATTI, M. L.; FARNSWORTH, N. R. The potential of alkaloids in drug discovery. **Phytother Rev**. v.15, n.1, p. 183-205, 2001.

CORNER, E. J. H. **The seeds of dicotyledons**. Cambridge: Cambridge University, 1976. v. 1, 311 p.

CORRÊA, M. P. **Dicionârio das plantas ùteis do Brasil e exóticas cultivadas**. Rio de Janeiro: Ministry of Agriculture, 1926. v.1., 747p.

DIAS, A. M.; MORAIS, M. C. **Morphoanatomical study and phytochemical prospection of the leaves of Caryocar brasiliense Cambess (Caryocaraceae) occurring in the Anâpolis - GO area**. 2011. ... f. Course Conclusion Work (Specialization) - Universidade Estadual de Goiàs, Anâpolis, 2011.

DINIZ, D. M. **Anti-inflammatory activity of microemulsion containing pequi oil (*Caryocar brasiliense*).**2015. Course Conclusion Work (Graduation) - State University of Paraiba, Campina Grande, 2015.

FONSECA, J. J. S. **Metodologia da pesquisa cientifica**. Fortaleza: UEC, 2002. Handout.

GERMANO, J. N. **Frutos do** cerrado. www.todafruta.com.br www.uraonline.com.br/culinariagoiana.html, en.wikipedia.org/wiki/Pequi; Accessed on: 07/09/2016.

GERMANO, J. N., SILVA, R. L. A.; SANTOS, E.M. Estudo Etnobotânico das plantas medicinais do cerrado do estado de Mato Grosso. **Brazilian Journal of Medicinal Plants**. V.3, n.1, pag. 23-31, 2007.

GIL, A. C. **Como elaborar projetos de pesquisa**. 5th ed. Sao Paulo: Atlas, 2010.

GIL, A. C. **Métodos e técnicas de pesquisa social.** 6th ed. Sao Paulo: Atlas, 2012.

HATANAKA, E.; CURI, R. Fattyacidsandwoundhealing: a review. **Revista Brasileira de Farmâcia**, v.88, n.2, p.53-58, 2007.

HENRIQUES, A. T.; KERBER, V. A. **Alkaloids: generalities and basic aspects.** In: SIMOES, C. M. O. et al. Farmacognosia da planta ao medicamento. 3. ed. Porto Alegre: UFRGS, 2001. p. 651-666.

HERZOG-SOARES, J. D. et al. In vivo trypanocidal activity of Stryphnodendron adstringens (barbatimao verdadeiro) and *Caryocar brasiliense* (pequi). **RevBras de Farmacogn**. v. 12, n.1, p. 1-2, 2002.

KERR, W. E.; SILVA, F. R.; TCHUCARRAMAE, B. Pequi (Caryocar Brasiliense Camb.) Preliminary information on a pequi without thorns in the stone. **Revista Brasileira de Fruticultura, Jaboticabal**, v. 29, n. 1, p. 169-171, 2007.

LAKATOS, E. M.; MARCONI, M. A. **Fundamentos de metodologia cientifica**. 6th ed. Sâo Paulo: Atlas, 2005.

LIMA, A. **Chemical characterization, evaluation of *in vitro* and *in vivo* antioxidant activity, and identification of phenolic compounds present in pequi *(Caryocar brasiliense* Camb.)** 2008. Thesis (PhD) - Faculty of Pharmaceutical Sciences, University of São Paulo, São Paulo, 2008.

LORENZI, H. **Arvores brasileiras: manual de identificaçâo e cultivo de plantas arbóreas nativas do Brasil**. Nova Odessa: Instituto Plantarum, 2002. p. 368.

LORENZI, H. **Arvores brasileiras: manual de identificaçâo e cultivo de plantas arbóreas nativas do Brasil.**Nova Odessa: Plantarum, 2000. v. 1.

LUCA, V.; PIERRE, B. S. The cellanddevelopmentalbiologyofalkaloidbiosynthesis. TrendsPlantSci. v. 5, n. 1, p. 168-173, 2000.

MANDELBAUM, S.H.; DI SANTIS, E.P.; MANDELBAUM, M.H.S.A. Cicatrisation: currentandauxiliaryresources-Part 1.**Anais Brasileiros de Dermatologia**, v.78, n.4, p.393-410, 2003.

MAGID, A. et al. Triterpenoidsaponinsfromthefruitof Caryocarglabrum. J. Nat. Prod. v. 69, n.1, p.196-205, 2006.

MARQUES, M. C. S.; CARDOSO, M. G.; GAVILANES, M. L. **Phytochemical**

analysis of the leaves and floral buds of Caryocar brasilienseCamb. In: SEMINARIO DO CENTRO OESTE DE PLANTAS MEDICINAIS, 1.2000, Rio Verde. Proceedings... Rio Verde: FESURU, 2000.

MARQUES, S.R.; PEIXOTO, C.A.; MESSIAS, J.B.; AL PEIXOTO, C.A.; MESSIAS, J.B.; ALBUQUERQUE, A.R.; SILVA JÛNIOR, V.A. The effectsof topical applicationofsunflower-seedoilon open wound in lambs. **Acta Cirùrgica Brasileira**, v.19, n.3, p.91103, 2004.

MELLO, J. P. C. et al. **Farmacognosia: da planta ao medicamento**. 3. ed. Porto Alegre: UFSC, 2001.

MIGUEL, M. P. **Neuroprotective action of the ethanolic extract of pequi bark in rat brains submitted to ischemia and reperfusion**. 2011. ... f. Thesis (Doctorate) - School of Veterinary and Zootechnics, Federal University of Goiás, Goiânia, 2011.

MOURA, L. R.; MARTINS, A. C.; VAZ, L. A. R.; ORPINELLI, S. R. T.; SILVA, T. L.; SANTOS, F. C. S.; MOURA, M. B. D. Hydroalcoholic extract of pequi bark (Caryocar brasiliense) in rats submitted to doxorubicin application. **Revista Ciência Rural**, v.43, n.1, jan, 2013.

OLIVEIRA, B. H. **Obtaining new drugs through biotransformation of natural products.** In: YUNES, R. A.; CHECHINEL- FILHO, V. Chemistry of natural products, new drugs and modern pharmacognosy. Itajai: UNIVALE, 2007.

OLIVEIRA, M. M.; GILBERT, B.; MORS, W. B. **Triterpenes in Caryocar brasiliense**. In: ANAIS DA ACADEMIA BRASILEIRA DE CIÊNCIAS, 1968, Rio de Janeiro. Proceedings... Rio de Janeiro: academy, 1968. p. 451- 452.

PAULA-JUNIOR, W. et al. Leishmanicidal, bactericidal and antioxidant activities of the hydroethanolic extract of the leaves of Caryocar brasiliense Cambess.

RevBrasFarmacogn. v. 16, n.1, p. 625-630, 2006.

PASSOS, X. S.; COSTA M.; SOUZA, L. K. H.; MIRANDA, A. T. B.; LEMOS, A. A.; FERRI, P. H.; SANTOS, S. C.; SILVA, M. R. **Antifung al activity of Caryocar brasiliensis against Paracoccidioides brasiliensis and Histoplasma capsulatum**. Proceedings of the XXI Brazilian Congress of Microbiology. Foz do Iguaçu, p. 60-71, 2001.

PENHA, A. R. S. **Study of the anti-ulcer activity of plants from the Araripe Plateau**. Monograph (Graduation in Biological Sciences), 2007. Universidade Regional do Cariri, Crato, 2007.

PIANOVSKI, A. R.; VILELA, A. F. G.; SILVA, A. A. S.; LIMA, C.G.; SILVA, K. K.; CARVALHO, V. F.M.; MUSIS, C.R.; MACHADO, S. R. P.; FERRARI, M. Use of pequi oil (Caryocar brasiliensis) in cosmetic emulsions: development and evaluation of physical stability. **Revista Brasileira de Ciências Farmacêuticas**, v.44, n.2, p.249-259, 2008.

QUIRINO, G. S. HealingpotentialofPequi (*Caryocar coriaceum* Wittm,) fruitpulpoil. **PhytochemLett.** v. 2, n.1, p. 179-183, 2009.

RIBEIRO, A. E. Space, man and his destiny in the north of Minas Gerais. In: FEDERAL UNIVERSITY OF LAVRAS. Department of Administration and Economics. Sustainable management of the cerrado for multiple use: **agroecology and development subproject**. Lavras, 1996. p. 11-18.

ROESLER, R.; CATHARINO, R. R., MALTA, L. G., EBERLIN, M. N., PASTORE, G. Antioxidant activityof Caryocar brasiliense (pequi) and characterization of components by electrospray ionization mass spectrometry.

Food Chemistry. London, v. 110, p. 711-717, 2θO8.

ROESLER, R.; MALTA, L. G.; CARRASCO, L. C.; HOLANDA, R. B.; SOUSA, C. A. S.; PASTORE, G. M. Antioxidant activity of cerrado fruits.

Ciência e Tecnologia de Alimentos, Campinas, v. 27, n. 1, p. 53-60, 2007.

SANTIAGO, F. M. **Study of the lectinic, toxic and hemolytic properties of the fruit of *Caryocar brasiliensis*.** Master's dissertation. Department of Biological Sciences. Uberlândia, Federal University of Uberlândia: 72. 1998.

SANTOS, M. S.; CAVALCANTI, D. S. P. **The medicinal use of pequi (Caryocar brasiliense).**2010. 3p. Alfredo Nasser College Union. Institute of Health Sciences. Aparecida de Goiânia, Goiàs, 2010.

SANTOS, F. S.; SANTOS, R. F.; DIAS, P. P.; JR., L. A. Z.; TOMASSONI, F. The Pequi crop (Caryocar brasiliense Camb.). State University of Western Paraná - Postgraduate Program in Energy in Agriculture - Master's Level. **Rev. Actalguazu**, Cascavel, v.2, n.3, p. 46-57, 2013.

SILVA, A. L. A. A.; DIAS, J. A.; FIGUEIRINHA, M. O.; **Benefits of Brazilian pequi (*Caryocar brasiliensis*), a fruit native to the cerrado regions of the center-west**. Dissertation. Integrated Colleges of Três Lagoas - AEMS. Três Lagoas. 2010.

SOUZA, Eli Regina Barboza de; FERNANDES, Eliana Paula; NAVES, Ronaldo

Veloso. **Physical and chemical characterization of pequi tree fruits (*Caryocar brasiliense Camb.*) from its regions in the state of Goiás, Brazil.** PesqAgropecTrop, v.37, n. 2, p. 93-99, jun/ 2007. Available at: <http://revistas.ufg.br/index.php/pat/article/viewFile/1833/3097>. Accessed on: September 07, 2016.

TONOZI-REIS, M. F. C. **Metodologia da pesquisa**. Curitiba: IESDE, 2006.

VERA, R. **Physical characterization of pequi tree fruits (Caryocar brasiliense Camb.) in the state of Goiás**. PAT. v. 35, n. 2, p. 71-79, 2005.

VIEIRA, R. F., MARTINS, M. V. M. Genetic resources of medicinal plants from the cerrado: a compilation of data. **Brazilian Journal of Medicinal Plants**. Botucatu, v. 3, n.1, p.13-36, 2000.

Printed by Books on Demand GmbH, Norderstedt / Germany